Cyber Explorers
A Kid's Guide to Being Smart, Safe and Cool Online

Welcome, Cyber Explorer!

The internet is a giant playground — full of games, apps and amazing tools that help us learn, create and connect with friends all over the world. But just like in real life, there are rules, challenges and sneaky "villains" we need to watch out for!

In this book, you will discover how to:
✦Protect your personal information like a secret treasure.
✦Outsmart cyber-villains by using your digital superpowers.
✦Explore the amazing world of Artificial Intelligence (AI).
✦Be kind online and help make the internet a safe, happy place.
✦Discover the fun side of cybersecurity and even future careers!

With colorful illustrations, fun activities, quizzes and creative projects, Cyber Explorers makes learning about the digital world exciting and easy.

Grab your pencils, imagination, and curiosity — it's time to become a Certified Cyber Explorer!

Written, Designed and Published in the USA

📖 TABLE OF CONTENTS

Welcome to the Cyber Playground

Internet- The Cyber Playground!

The Internet is like a giant playground where billions of people work, play, learn, and interact. But instead of swings and slides, you have apps, games, videos and so much more!

- Every time you watch YouTube, scroll through Instagram, see a movie on Netflix, search on Google, or play online games—you are on the Internet!
- Think of Internet as a 'giant machine' that connects the whole world through computer applications, software and networks.
- And just like your school playground has a fence to keep you safe, the Internet needs a 'virtual fence' too. These fences are things like passwords, face ID, or other safety tools that protect you while you play and explore online."

FACT The internet is so big that if it were a country, it would be the largest in the world! 🌍

G L O S S A R Y

HACKER the person seen in videos wearing a hoody he or she tries to break into systems causing bad things. But some are 'good hackers' who help keep things safe (called ethical hacker)

FIREWALL A 'digital wall' that protects computer and systems from intruders who tries to sneak in to systems.

PASSWORD YOUR secret key to unlock phone or an App.

CLOUD This is not a cloud in the sky! In fact, they are data centers which are large physical facilities with huge numbers of computers—creating a giant virtual space where the apps you use run and where your photos, videos and files are stored.

DATA Data is the nerdy name for 'information' which could be a picture, video you watch, online chat with a friend.

These are some of the "building blocks" of
Cyber Playground!

YOUR TURN!

Instructions for Kids:
Imagine the internet as a playground just for you.
What would the swings, slides or fences look like?
Use this space to draw your own Cyber Playground!

CHALLENGE Extra Challenge: Add labels like "Apps," "Games," or "Firewall" to parts of your drawing.

Reflection Question:
If your playground had no fence, what could happen?

Protecting Your Digital Self

So now you know that 'information' or 'data' is part of the Internet. Just like you would protect a treasure chest, you need to protect your information online so it doesn't end up in the wrong hands.

You need to Keep it locked up and private.
Don't give it away to strangers or post it online.

Private Treasure (KEEP SAFE!)
Name
Birthday
Address & phone number
Passwords
School name

Public Fun Facts (SAFE TO SHARE!)
Favorite color 🖍️
Favorite food 🍕
Favorite sport ⚽
Favorite cartoon 🦆

👍 Rule of Thumb: If you would not write it on a giant billboard for the world to see, do not share it online!

Use a long, strong password.

PRIVATE vs. PUBLIC

PRIVATE | PUBLIC

PASSWORDS

Strong passwords keep your accounts safe.

Make it hard for others to guess.

12345

Keep your password safe.

Passwords (remember the fence) are your digital superheroes. The stronger they are, the harder it is for villains to break in.

Weak Passwords (too easy to guess)
12345
password
ilovepizza

Strong Passwords (superhero level strength):
Use UPPER and lower case letters.
Add numbers and symbols.
Examples:
Weak → sunshine
Strong → SuNSh!n3#2025

Make it unique—do not copy anyone else's passwords.

Activity: Password Recipe

RECIPE

INGREDIENTS	INSTRUCTIONS
- - - - - - - - -	- - - - - - - - -
- - - - - - - - -	- - - - - - - - -
- - - - - - - - -	- - - - - - - - -
- - - - - - - - -	- - - - - - - - -

Create your own strong password using this recipe:

🐕 Your favorite animal

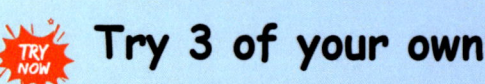 A number you like

🍀 A symbol !@#$%

ABC A CAPITAL letter

Example: Tiger#42!

💥 **Try 3 of your own!**

🧠 **Reflection Question:**
Why do you think 12345 is a bad password?

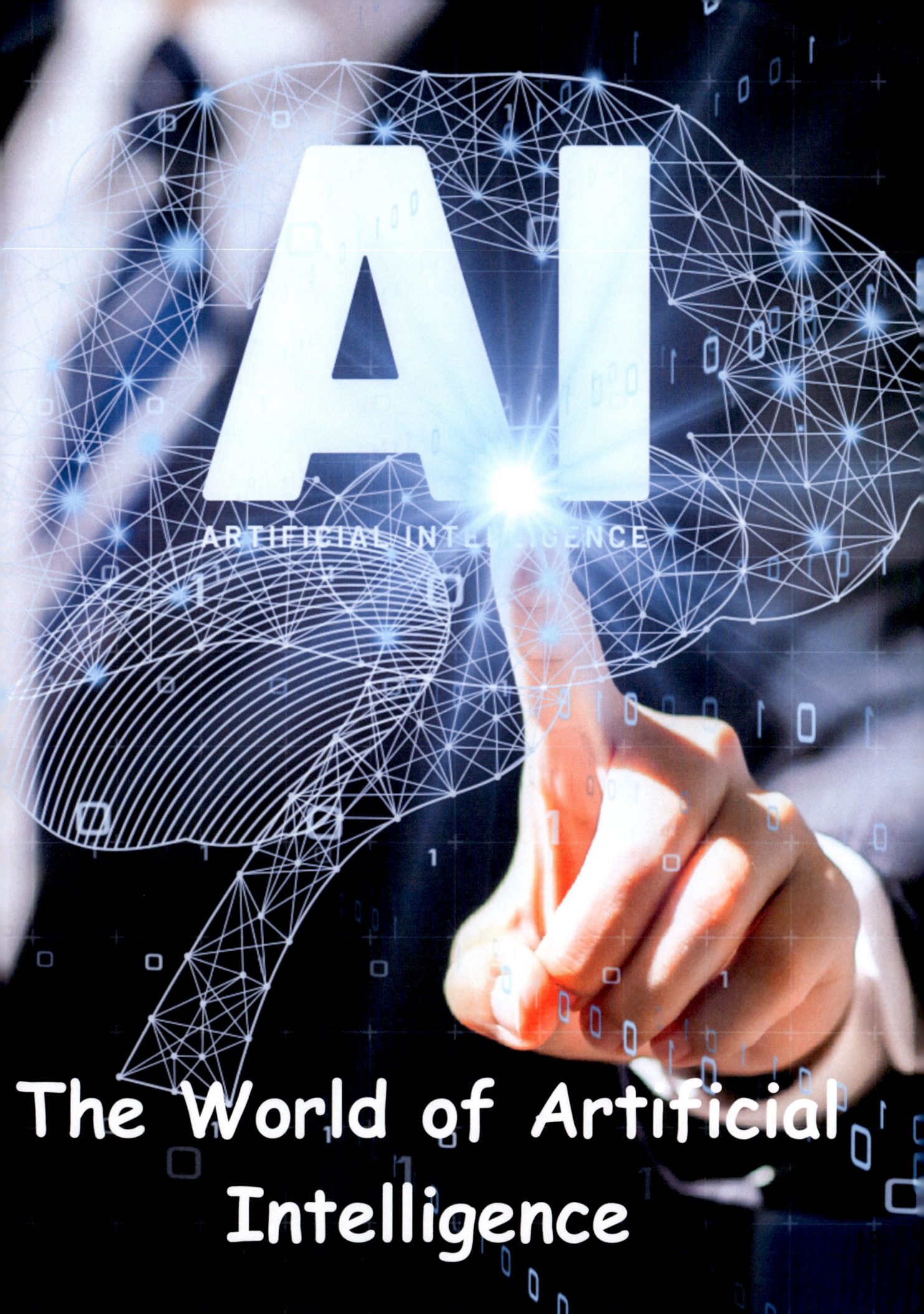

Artificial Intelligence, or its acronym AI, is like giving a computer a SUPER brain!
It can learn patterns and make decisions.
It's used in games, robots, cars and even in your favorite apps.
AI is not magic—it's smart technology created by people.
AI is like having a super-smart pet robot—it doesn't know everything, but it can learn tricks if you train it.

Fun Fact: When you use voice assistants like Alexa or Siri, you are talking to AI! 🎤🤖

WHERE DO WE SEE AI?

Self-driving car

Robot vacuum

Game opponent

Netflix/YouTube suggestions

Face filter

Smart speaker

AI is everywhere around us!
It helps cars drive, cleans our homes and makes games fun.
It suggests new songs, videos and shows.
It even puts silly filters on your face!

Smart Helpers: Siri, Alexa, Google Assistant 📢
Smart Games: Enemies that get harder as you play 🎮
Smart Cameras: Phones that recognize your face 📱
Smart Choices: Netflix or YouTube suggestions 💻
Smart Cars: Self-driving or cars that beep if you drift 🚗

LET'S DRAW!

 **Circle the things that use AI**

🧑‍🦱💭⭐ **Reflection Question:**
Do you think AI can ever be smarter than people?
Why or why not?

(Answers: Robot vacuum, YouTube suggestions, traffic light)

Future with Artificial
Intelligence

AI learns from data (like how kids learn from books, teachers and experiences).
If it sees many examples, it gets smarter — but if the examples are unfair or wrong, it can make mistakes.

Example: If an AI only sees pictures of bananas, 🍌it may think apples 🍎 don't exist.

AI Superpowers (and Limits)

Superpowers

- **Pattern Spotter**👀 – AI can look at millions of pictures and say "dog!" or "stop signs" super fast.
- **Language Helper**🌍 – Translates words so people from different countries can talk.
- **Car Driver**🚗 – Helps cars see traffic lights and roads.
- **Doctor's Assistant**🧑‍⚕️📋 – Scans X-rays to detect sickness early.
- **Game Player** 🎮 – Plays chess or video games and sometimes wins against humans!

Limits

- **No Real Feelings**💔 – AI can not truly feel happy or sad.
- **Gets Confused** 🤷 – A purple cow might totally trick it!
- **Needs Training** 📚 – AI only knows what it's shown.
- **Makes Mistakes** ❌ – Wrong data = wrong answers.

19

Good vs. Bad Uses of AI ⚖️

Good Uses: Helping with homework, discovering planets, curing diseases. ✨

Bad Uses: Making fake videos, hacking, spreading lies. 🚫

Design Your Own Helpful Robot

Draw your own AI robot that can help
you in daily life.
"My robot helps me with _____."
"It looks like _____."
"Its special power is _____."

CYBER VILLAIN LINEUP!

The internet is full of fun, but sometimes sneaky troublemakers try to cause problems. These villains don't live in video games—they try to sneak into real computers! Let's meet them.

VIRUS MONSTER

PHISHY CAT

HACKER FOX

Villain 1: Virus Monster - A slimy green blob with sharp teeth. Sneaks into your computer to break files, slow things down, or make games crash. By hiding in downloads, strange links, or infected apps. **Danger Level: High!**

Villain 2: Phishy Cat- A sneaky cat with a fishing pole, dangling fake candy or emails.
Tricks you into clicking bad links or giving away secrets. Sends fake messages that look real. Pretends to be your bank, teacher or even a game site. **Danger Level: Tricky!**

Villain 3: Hacker Fox - A clever orange fox wearing a mask. Tries to sneak past your password and steal your treasure (your data!). Guesses weak passwords like "12345." Pretends to be you online. **Danger Level: Sneaky!**

HOW TO DEFEAT THE CYBER VILLAINS!

Every villain has a weakness. Just like in video games, you can use the right tools to beat them! Lets's see how.

VIRUS MONSTER

Virus Monster's Trick: It sneaks into computers through. downloads, bad links.

Antivirus Shield

Gab! shield protects my computer & deletes bugs

PHISHY CAT

Phishy Cat's Trick Sends fake emalls or messages to trick you into clicking.

Hai I caught your fake email!

HACKER FOXS TRICK

Hacker Fox's Trick Guesses weak passwords like.,1245 or 'password."

Fun Tip:
Just like in video games.

Virus Monster
Your Hero Tool: Antivirus Shield 🛡️
Updates protect your computer. Software scans and deletes bugs.

Phishy Cat
Your Hero Tool: Detective Glass 🔍
Look closely at links before clicking. Ask an adult if something feels fishy.

Hacker Fox
Your Hero Tool: Password Hero Key 🔑
Use long, strong, unique passwords. Mix letters, numbers, and symbols.

Circle the hero tool you would use if:

- Someone sends you a weird link → _____

- Your game starts acting slow with bugs → _____

- A hacker tries to guess your password → _____

"I know how to beat the villains!"

CREATE & BATTLE A CYBER VILLAIN!

Draw your villain inside the box below.

Give it wild colors, shapes or funny features.

CHOOSE YOUR HERO TOOL

CHOOSE YOUR HERO TOOL

VILLAIN'S NAME:

SUPERPOWER:

WEAKNESS:

VILLAIN SAYS:

HERO REPLIES:

I DEFEATED MY CYBER VILLAIN!

I AM NOW A CYBER HERO!

Cyber Tools = Digital Superpowers

Encryption = Secret Codes

Encryption is like writing a message in a secret code so only the right person can read it. Hackers who try to peek will just see scrambled letters!

Example: HELLO →
IFMMP

Firewall = Digital Fence

A firewall is like a big fence around your computer. It decides what type of information can come in and what must stay out. It makes it difficult for bad guys to cause havoc! Imagine your house has a tall fence. Friends with a secret password can come inside, but strangers stay out.

Example:
Good traffic (like your learning App) → ✅ allowed in.
Bad traffic (like a hacker's bug) → ❌ blocked at the fence.

Two-Factor Auth = Double Lock

Sometimes, one lock isn't enough.
Two-Factor Authentication has two locks
on your treasure chest:

1. Password
2. Text code

Even if a villain guesses your password, they can not
get in without the second lock!

Example:
Step 1: Type your password.
Step 2: Get a secret code texted to your phone.
Step 3: Enter the code → 🎉 Now you're in!

Become a Secret Code Master!

Shift each letter in a word forward by 2 letters in the alphabet.
Example: CAT → ECV
Example: DOG → FQI
Now write your own secret message here:

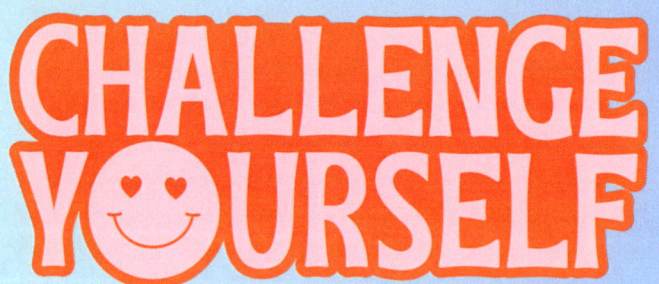

1. Encode your name.
2. Encode your favorite food.
3. Pass the code to a friend and see if they can crack it!

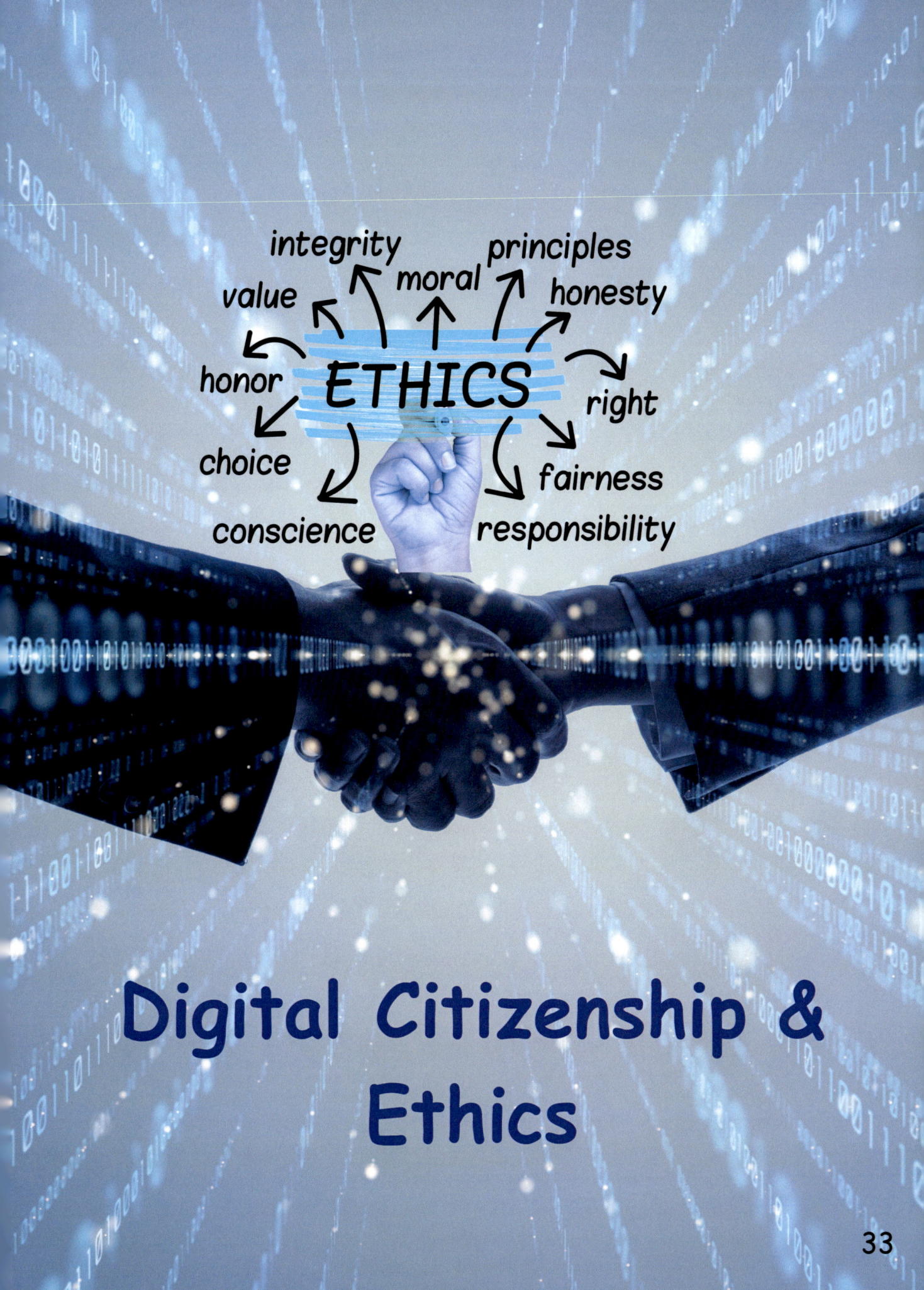

Digital Citizenship & Ethics

Be Kind Online

Treat people online the same as in real life.

Cyberbullying hurting with words or pictures.

MEAN WORDS

The internet is a shared space like a playground. Just like in real life, we need to be kind, respectful and fair to others.

GOLDEN RULE Treat people online the same way you would treat them face-to-face.

STOP BULLYING Cyberbullying = Hurting with words or pictures. Examples: mean comments, sharing embarrassing photos, excluding people from group chats.

Kindness = Power! Positive words spread faster than negative ones.

WHAT WOULD YOU DO?

A stranger asks for your photo	A classmate is being bullied	You see a scary video

Here are some real-life choices you may face online and you must pick or write what they would do.

A stranger asks for your photo.
Best choice: ❌ Don't share. Tell a parent/teacher.

A classmate is being bullied online.
Best choice: ✅ Speak up, be supportive and tell an adult.

You see a scary video
Best choice: ✅ Stop watching, talk to a parent.

AI and Ethics

Can robots be unfair? Yes –
if they learn from biased data.

AI is powerful, but it can also make mistakes if it only learns from unfair or incomplete data.

Fairness Matters:
If a robot only sees cats, 🐱 it may think dogs 🐶 do not exist.

If an AI game only shows boys as heroes, it may forget girls can be heroes too.

We must teach AI to be fair, just like we teach kids to share and include others.

Your Turn!

Draw a "Cyber Code of Conduct" poster.

Rules to include:
Be kind, not mean.
Never share private info.
Think before you click.
Respect everyone online.
Tell an adult if something feels wrong.

CYBER FUN
TIME

Privacy Shield Project

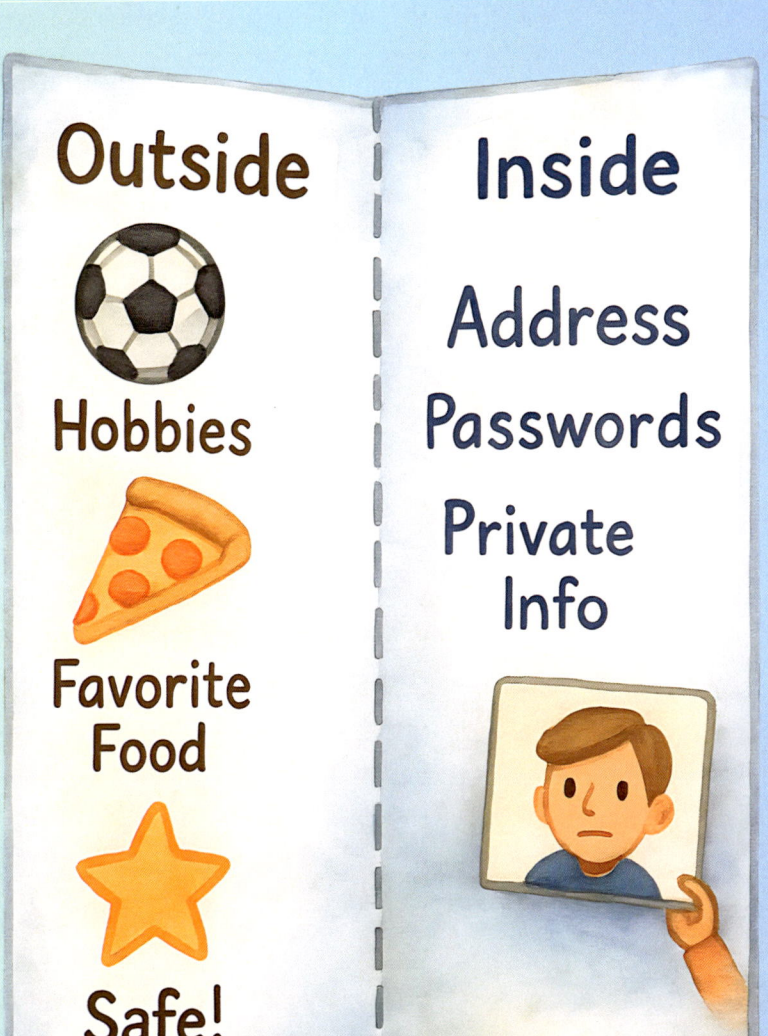

Fold a paper in half.

Inside (Hidden): Write/draw private info (address, phone number, passwords, school).

Outside (Public): Draw/write safe info (favorite food, hobbies ⚽, pets, favorite color 🎨)

Lesson teaches the difference between what should stay private and what's okay to share

Detect Fake News

Circle ✅ for Real or ❌ for Fake:

- NASA launches a new satellite into space.
- Aliens open a McDonald's on the Moon.
- Scientists discover a new species of frog in the Amazon Rainforest.
- Unicorns are found living in New York City parks.
- Doctors use robots to help with surgery.
- Self-driving cars are being tested on real roads.
- Chocolate milk comes from brown cows.

Fake vs. Real News Quiz
Answers

1. NASA launches a new satellite into space.

2. Aliens open a new McDonald's on the Moon.

3. Scientists discover a new species of frog in the Amazon Rainforest.

4. Unicorns are found living in New York City parks.

5. Doctors use robots to help with surgery.

7. Self-driving cars are being tested on real roads.

Chocolate milk comes from brown cows.

Mini Lab – Secret Agent Badge

- Draw a badge on paper or card.
- Add a superhero name (like "Cyber Ninja" or "Password Panther").
- Write 3 personal Cyber Rules (e.g., I will not share my password, I will be kind online, I will double-check links).

Jobs of Tomorrow

- Ethical Hacker – Finds problems in systems before the "bad guys" do.
- AI Designer – Builds robots and smart apps to help people.
- Cyber Detective – Tracks down digital crimes.
- Game Security Tester – Makes sure games are safe and fair.

Which Job Fits You?

- Do you like solving puzzles? → You'd make a great Hacker.
- Do you like drawing or creating new things? → AI Designer.
- Do you like helping others? → Cyber Detective or Teacher.
- Do you love gaming? → Game Security Tester.

WHICH JOB FITS YOU?

| Like puzzles → Ethical Hacker | Like drawing → Designer | Like helping → Detective | Like games → Tester |

Internet Timeline

Past (1990s): Dial-up internet, slow but new.
Present: Social media, smartphones, AI everywhere.
Future: Quantum computers, space internet, AI astronauts.

Cyber Explorer's Final Pledge

"I promise to protect my information, be kind online, think before I click, and use technology to make the world better."

CERTIFIED CYBER EXPLORER!

I promise to protect my information, be kind online, think before I click, and use technology to make the world better.

CERTIFIED CYBER EXPLOER

Joby James

Joby James has built an illustrious career at the intersection of engineering, cybersecurity sales and tech entrepreneurship. As the founder of the cybersecurity firm Cybersainya, his team helps companies safeguard data while equipping organizations to navigate the digital world safely and responsibly.

Before becoming an entrepreneur, he served as a cybersecurity sales leader at Cisco, where he helped government agencies, K-12 schools, universities, and hospitals build effective cyber resilience. Joby believes that safeguarding and mentoring young minds is more critical now than ever. He enjoys creating programs and content that make it easier and safer for young people to navigate the digital world.

Dr. Diya Abraham, Ph.D

A Researcher at heart and mentor by calling, Dr. Diya Abraham earned her doctorate in Neuroscience from the Max Planck Institute in Germany and completed a post-doctoral fellowship at the University of California, San Francisco. Her work on circadian rhythms, genes that keep our inner clocks in sync—has appeared in leading peer-reviewed journals and has been showcased on international conference stages.

Through Bee Little Curious, the company she founded to enrich curious minds, she creates evidence-based educational resources that blend science with creativity. She also partners with schools to embed STEM principles into engaging curricula designed to inspire the next generation of learners.

www.ingramcontent.com/pod-product-compliance
Lightning Source LLC
Chambersburg PA
CBRC090828120626
46547CB00008B/627